Legal information

© 2020

Johannes Wild
Lohmaierstr. 7
94405 Landau
Germany

E-Mail: 3dtechworkshop@gmail.com
Phone: +4915257887206

This work is protected by copyright. Any use outside the narrow limits of copyright law without the author's consent is prohibited. This applies in particular to electronic or other reproduction, translation, distribution, and making publicly available. All rights reserved.

Thank you!

Thank you so much for choosing this book! With this book, you can develop a solid knowledge about 3D printing, and you will learn everything you need to know in terms of FDM 3D printing.

This guide will show you in particular how to prepare (i.e. slice) and print objects by using a 3D printer!

Table of Contents

Legal information 1

Table of Contents... 2

1 Motivation 4

2 Introduction to 3D Printing 5

2.1 3D Printing Technologies and Procedures.. 5

2.2 3D Printer Purchase Advice 8

2.3 Essential 3D Printing Equipment 11

2.4 3D Printer Setup and first Print 12

3 Introduction to 3D Printing Software 16

3.1 Slicing Software and File Formats... 16

3.2 Further 3D Printing Software 18

3.3 Introduction to Cura and its important Settings 23

4 Advanced Settings and further example Objects 30

4.1 Upgrade Parts for the CR-10 3D Printer 30

4.2 Slicing and printing of more complex Objects... 35

4.2.1 Low-Poly-Dog Sculpture 35

4.2.2 Honeycomb Vase 39

4.2.3 Various Filling Patterns... 41

4.2.4 Further Tips for Slicing.. 43

5 Filaments and Materials.. 45

6 3D Printer Maintenance and Troubleshooting... 48

6.1 Repair, Maintenance, and Troubleshooting... 48

6.2 Troubleshooting Guide 50

6.2.1 Poor Adhesion.. 50

6.2.2 Under-Extrusion and Over-Extrusion 50

6.2.3 Stringing or Oozing 51

6.2.4 Layer Separation 51
6.2.5 Clogged Nozzle 51
6.2.6 Poor Infill 52
6.2.7 Vibrations and Ringing 52
6.2.8 Warping 52
6.2.9 Scars on the Print 53
7 3D Scanning 54

1 Motivation

I'm sure you've already heard some fantastic stories about 3D printing and seen some fascinating photos of 3D printed objects that have sparked your enthusiasm for this subject. If not, then let's get in touch with this ingenious technology. I'm sure that you'll be amazed at what's possible in the world of 3D printing. In this book, you will be guided step by step, and you will learn how to use an FDM 3D printer and its software through the expertise of an engineer and experienced 3D printer user.

Everything you need to know to get started in 3D printing is explained in detail. You will learn how to print objects (downloadable for free) and how to materialize your projects, ideas, or prototypes. This book is made for all technically open-minded people who are interested in 3D printing.

Ideally, you should buy a 3D printer. But don't worry, you don't have to pay a large amount of money for it. High-quality 3D printers using FDM technology are available for less than $ 500.

If you don't want to buy a 3D printer, you will still benefit from this book. On the one hand, through knowledge of the most important processes and principles of 3D printing technology, on the other hand, utilizing the possibility to use a so-called makerspace or an external 3D printing service provider.

A makerspace is a place where several and different technical devices such as 3D printers, laser scanners or laser cutters are available to members for free use. These rooms, which are equipped like a workshop, offer the necessary space and the right equipment for the realization of your projects. You can easily find out online where the next makerspace is located.

And before we start with the basics of 3D printing, we want to take a look at a selection of 3D printed objects first (Fig. 1).

Figure 1: 3D printed objects: vase, sculptures, and toys

With the help of this book, you will be able to print many more great objects in the future. But you have surely been able to pick up some inspiration. And that's why we're now taking a closer look at the basic principles of 3D printing in order to get started with printing as soon as possible.

2 Introduction to 3D Printing

2.1 3D Printing Technologies and Procedures

The mainly presented and discussed 3D printing technology in this workshop is called FDM or FFF. FDM is the short form of "Fused Deposition Modeling". The heart of a FDM 3D printer is a nozzle equipped with a heating element, the so-called "Hot-End" (see Fig. 2). This part melts filament - the printing material - and deposits it, layer by layer on the printing bed or the previously printed layer. In this way, the print object gains height layer by layer.

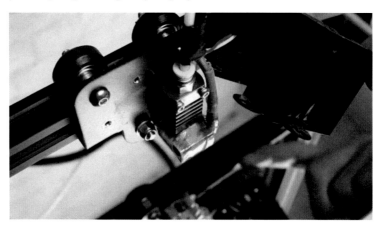

Figure 2: Hot-End of a 3D printer (Creality CR-10)

A plastic filament is usually used as a printing material. PLA is perhaps one of the easiest and common materials when it comes to 3D printing.

PLA is a plastic that is obtained from regenerative sources such as corn starch and is therefore biocompatible and food safe. Other plastics such as ABS are more difficult to print but have better mechanical properties. The layer thickness is usually chosen in the range between 0.1 mm and 0.4 mm, whereby the print quality increases with smaller layer thickness. However, the smaller the layer thickness, the longer the printing time.

To ensure that the print adheres well to the printing bed and does not come loose during printing, a heated printing bed in the range of approx. 60° Celsius is recommended. Some 3D printing enthusiasts use all kinds of additional equipment such as adhesive tapes, glue sticks, hairspray, or permanent printing plates to achieve good printing bed adhesion. However, with well-coordinated settings and a calibrated printing bed, it normally works without those kinds of stuff. But there is an insider tip regarding the printing bed:

A mirror is used as a printing bed to provide a perfectly flat surface for the print object. The reason for using a mirror is that mirrors have to be extremely flat to provide a distortion-free reflection.

The print head can move through all three spatial directions by means of so-called stepper motors. This makes it possible to print almost any shape (see Fig. 3).

Figure 3: One out of many stepper motors mounted on a 3D printer

In addition to FDM technology, there are also other 3D printing procedures, such as selective laser sintering and stereolithography, SLS and SLA for short. These are mainly used in industrial sectors. A laser is used to fuse a plastic powder, liquid resin, or metal powder into solid components. 3D printers from manufacturers such as Formlabs, Sintratec, or EOS, use these technologies. These 3D printers are generally somewhat more complex and expensive than those using the FDM method and, therefore, only briefly mentioned in this book.

Back to our FDM printer: Now, the filament, which is rolled up in a wire-like structure on a spool, must, of course, first reach the head and the nozzle of the printer. This is done by transport via another stepper motor, which grips the filament with a gear wheel and moves it towards or away from the nozzle (see Fig. 4). Fans attached to the hot-end provide temperature control on the one hand and cooling of the liquefied plastic on the other and thus solidification in the printed form.

The printer presented in this book has only one nozzle, so only one color can be used in a printing job unless a mixed-color filament is used.

However, there are also so-called "dual extruders," such as the Creality CR-X. "Dual Extruder" means the 3D printer is equipped with two nozzles, and thus two types of filament or two colors can be processed in one print object.

Figure 4: Stepper motor for transporting the filament from the spool to the nozzle

Before we start our first printing job, there is some advice on buying FDM printers in the following.

Legal notice: *All the following recommendations and interpretations reflect my personal opinions and experiences. I have no business relations with the following 3D printer or filament manufacturers. This also applies to all other products - including 3D printing software - presented in this book.*

2.2 3D Printer Purchase Advice

As mentioned in the introduction, high-quality 3D printers are available for less than $ 500. For example, the Creality CR-10 (see Fig. 5). The CR-10 can be ordered directly from China, which works well, but often takes a long time. Fortunately, there are now also cheap and good offers from other warehouses selling on Amazon, eBay, and co.

The CR-10 series is highly praised in the 3D printing community, and in my opinion, that is appropriate. The print quality is outstanding, and the price is relatively low, depending upon models with approx. $ 400 - $ 800. Another advantage is the large building space with a printing bed of 30 x 30 cm (11.81 x 11.811 inches) and a printing height of 40 cm (15,748 inches). Therefore this printer is my explicit purchase recommendation. 3D printing is explained in this tutorial by using this printer.

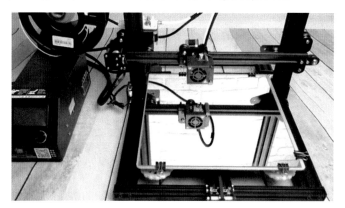

Figure 5: Creality CR-10 3D printer upgraded with a mirror as a printing bed surface

However, the basics presented in this book are broadly applicable to all FDM 3D printers, so if you choose a different one, that is not a problem. Meanwhile, there are also other versions of the original CR-10.

For instance, the CR-10S and the CR-10S Pro. These offer some useful upgrades like filament sensor, two motorized z-axes, automatic printing bed leveling, and other helpful modifications.

Furthermore, there is a CR-10 version with a printing space of 50 x 50 x 50 cm (19,685 x 19,685 x 19,685 inches).

Other recommended 3D printers are the devices of Makerbot. For the so-called replicator, for example, you have to pay about $ 2000. The Makerbot devices are also designed for hobby and semi-professional use.

Besides Makerbot, Ultimaker's 3D printer models are among the pioneers when it comes to 3D printing, and Ultimaker is one of the leading manufacturers in this field. These devices also achieve a very high print surface quality and are therefore listed as a second purchase recommendation. For an Ultimaker, you have to plan with approximately $ 2000 - $ 5000. From the Netherlands, the Builder Desktop 3D printer series with its "small," "medium," and "large" models are known, starting from approx. $ 2500. Another noteworthy FDM printer is the Dremel Idea Builder, priced with approx. $ 1500.

There are numerous further 3D printer manufacturers and models. As an orientation to the 3D printers available on the market, however, these recommendations should be sufficient.

As a conclusion, it can be said that the CR-10 series, with its newer models like the CR-10S and/or the CR-10S Pro offers a favorable and good entrance into 3D printing.

The Ultimaker as a second recommendation also offers a very good entry, although a somewhat more expensive one.

Now I mentioned in the beginning that if this is still too expensive for you, you don't even have to buy yourself an own 3D printer to realize your ideas. At www.3dhubs.com, for example, you can start a 3D printing job by uploading a 3D model and thus find a suitable 3D printers provider anywhere in the world. Thanks to the Internet, it is possible to start 3D printing jobs by using the service of private 3D printing owners from the neighborhood or commercial industrial providers. SLS and SLA methods are also often offered.

Alternatively, there are other service providers such as Protolabs (www.protoblabs.com), Shapeways (www.shapeways.com), and co.
At Protolabs, for example, you can upload your desired 3D model as a ".stl" file and get a direct price calculation, including design analysis.

Or you choose the possibility of using a makerspace.

2.3 Essential 3D Printing Equipment

But if you are enthusiastic about buying a 3D printer, the following question arises: What other equipment do you need in addition to the actual printer so that you can get started? On the one hand, a printing material, i.e., a filament. I recommend starting with PLA because it's the easiest and safest material to use in this context.

There are probably as many filament manufacturers as 3D printer manufacturers, and it is not always easy to keep track of this sector. That's why you'll find a separate chapter on filaments further on in this book.

I recommend starting with a PLA+ filament from the award-winning filament manufacturer Sunlu. Alternatively, you can buy PLA+ from the manufacturer Tianse as well (see Fig. 6). The "+" stands for further development of PLA through increased mechanical properties combined with equally good printability. Filaments of Sunlu impress with the surface quality of the prints and the low manufacturing tolerance of +/- 0.02 mm as a quality feature. Make sure you choose the right filament diameter for your printer. For the CR-10 series, you'll need 1.75 mm. However, there are also 3D printers that require a 3 mm filament diameter due to their nozzle size. Both diameters are usually offered. Needless to say that you can also choose from numerous colors.

Figure 6: 3D printing filaments of the manufacturers Sunlu and Tianse

In addition to filament, you should also get a mirror for the printing bed, a storage medium (memory card) for the print files, some cleaning agents for the printing bed and superglue. Optionally some model colors if you want to paint the printed parts later on and other useful tools like some pliers and a spatula to remove the finished parts from the printing platform. The CR-10 comes with a basic set of tools.

Figure 7: Additional tools and equipment for 3D printing

2.4 3D Printer Setup and first Print

Now it's time to finally get to work with your 3D printer and master the first print. The CR-10 has to be assembled first after delivery. Don't worry; it's done very fast and pretty easily.

All parts are already prefabricated and have been disassembled into three parts for shipping reasons only. The best way to proceed is to follow the manufacturer's video instructions. You can scan the QR code on the printer or the manual with your smartphone or view the manual in the Youtube channel of the manufacturer "Creality". You may not need to assemble other printers by yourself. Some printers, however, are available as assembly kits only and must be constructed from scratch.

After the printer has been assembled, we have to level the printing bed to complete its proper function (see Fig. 8). Newer 3D printer models already have a sensor for automatic adjustment of the distance from the nozzle to the printing bed. But here it also makes sense to carry out a manual calibration at the first start-up in order to obtain a well-leveled printing bed surface.

For the process of manual calibration, heat up the printing nozzle to the desired printing temperature, thus approximately 210° Celsius and the printing bed to approximately 60° Celsius manually via the menu of the 3D printer.

Please make sure that leveling is performed in this warm condition, as the materials expand by the influence of heat. Then select "Home Position" in the printer's menu "Prepare". The printer will now move to a defined home position.

In the case of the CR-10, this is the left front corner of the printing bed. An identical starting position is necessary to always achieve the same z-distance relative towards the printing bed at the beginning of a printing process.

Calibration (leveling) is performed best with a distance gauge or feeler gauge of 0.1 mm. If you do not have such a gauge at hand, you can alternatively use some paper sheets as well. To check the distance between nozzle and printing bed, place the distance gauge or the piece of paper between nozzle and printing bed. The distance gauge or the paper sheet should just fit between the nozzle and printing bed. You should be able to hear a scraping noise when using a paper sheet. Adjust the distance with the setting wheels of the printing bed if it is necessary.

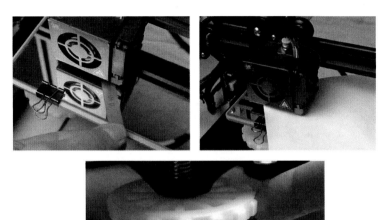

Figure 8: Printing bed leveling using a gauge or a piece of paper

Perform this procedure in all four corners and then repeat it one more time. Make sure that if you are using a piece of paper, the sheet can be hardly moved. Then the distance is exactly right.

If the first print does not adhere, reduce the distance between nozzle and printing bed slightly. A too-large distance between nozzle and printing bed is one of the most common beginner's mistake. However, the nozzle should not grind on the printing bed either when moving the print head.

If you use a 3D printer without a sensor, such as the first CR-10, this printing bed calibration should be performed periodically after every 10 - 20 prints. For printers with automatic distance sensors, such as the CR-10S Pro, this calibration is generally no longer necessary. You will also find helpful videos on the Youtube channels of the respective manufacturers for setting up the automatic distance regulation. Before we start our first printing job, a few safety instructions follow. To melt the filament, the printing nozzle has to heat up to approximately 200° Celsius. The printing bed also heats up to approx. 60° Celsius. Therefore there is a high risk of burns, especially when touching the nozzle. Do not touch moving parts either while the printer is running, as there is also a danger of crushing. Find a suitable location for the 3D printer, ideally a somewhat isolated room with a stable workplace, as the printer generates some noise and vibrations. Also, make sure that children and pets do not have unattended access to the printer. It is a good idea to wait a few minutes at the start of printing to check whether all operations are being carried out correctly. It is also recommended to check the printing procedure every now and then during printing. Now, all necessary preparations have been made, and everything important for the first printing is explained. Traditionally, the CR-10 series starts with a small waving Chinese lucky cat as a first print (see Fig. 9).

Figure 9: A Chinese lucky cat for the first printing job

With the newer models, the first print could also be a small dog or another object. As a filament, you can either use the test filament supplied with the 3D printer or a separately purchased one. Remove the SD card supplied with the 3D printer from the USB adapter and insert the SD card into the 3D printer without the adapter. Alternatively, you can connect the printer to a PC using a USB cable. Then switch on the 3D printer and navigate to "Print of SD Card," then start the file with the cat. If your 3D printer does not come with a suitable printing file, you can skip this chapter first or search for a file online (".gcode" file is required). Next, the 3D printer will start heating up the nozzle and the printing bed. You will see a rising temperature displayed on the menu. This process takes about 3 - 5 minutes. As soon as the desired temperatures are reached, the printing starts automatically. First, the head moves to the "HOME" position, and then the actual printing starts. Check whether the filament adheres evenly to the printing bed by looking at the printing bed and the nozzle. If not, stop printing and level the printing bed again. When the filament is evenly deposited on the printing bed, you can raise your feet and watch the 3D printer as it works. After approx. 2 - 3 h, the lucky cat is done. To remove the finished object from the printing platform, it is best to let the heating bed cool down completely, as the printed model still adheres very strongly to the printing bed when the heating bed is warm. After cooling down, use a metal spatula to get evenly under the object from all sides and lift it slightly.

After a little back and forth, it should come loose.

3 Introduction to 3D Printing Software

3.1 Slicing Software and File Formats

You have only been able to immediately print the lucky cat with your 3D printer because of the ".gcode" format of the file. By the way, you can see this in the file suffix.

Such a ".gcode" file has to be created first. Usually, 3D models can be downloaded as ".stl" files. Therefore we will proceed now to the next chapter in 3D printing. Let's take a look at the necessary software and file formats. By the way, you can download lots of 3D models for free at online platforms.

If you want to create your own components, you must use engineering software. But more about that later.

The already mentioned ".gcode" file format represents a machine code that provides the necessary movements and other printing information to the printer. This file is created using a so-called slicing program or "Slicer" for short.

"Slicer" because it divides the geometry of the desired object into layers. The software supplies the printer with the necessary movements in the spatial directions so that the object can be printed (see Fig. 10).

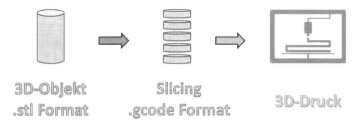

Figure 10: Necessary steps in 3D printing

As a slicing software, a selection of free and chargeable software is available. As a free software, for example, "Slic3r", which is written with the number 3 instead of an "e" is available. Cura, an open-source slicing software from the 3D printer manufacturer Ultimaker, can also be downloaded for free. If you want to buy slicing software, take a look at "Simplify3D". This software is priced at about $ 150.

My recommendation, however, is Cura. On the one hand, the program is free of charge; on the other hand, it is designed by one of the leading 3D printing manufacturers. And it's ideal for beginners.

The program works with almost all FDM 3D printers and impresses with its easy handling. Cura can be downloaded at the Ultimaker website (www.ultimaker.com).

If an object has been constructed with CAD construction software, it can be saved in the so-called ".stl" format. In this format, the geometric information of three-dimensional objects is optimally prepared for 3D printing.

You then import the ".stl" files into your slicing software and slice them into a machine code for your printer (".gcode").

There are many online platforms offering ".stl" files; most of the files are created by members of the 3D printing community and are often available for free for private, sometimes even for commercial use. One of the biggest platforms is Thingiverse (www.thingiverse.com). This platform offers free CAD files as ".stl" formats. You don't even have to register for it, and you can download the files right away.

There is, however, one thing to note:
The Creative Commons licenses of the files.

These licenses can be found on Thingiverse at approximately the left, middle page height of an object. With Creative Commons licenses, an author or creator of a work can easily grant the public rights to use his or her work.

There are different licenses. Depending on the license, only printing for private purposes is permitted or sharing, and modification of the work is explicitly desired, or even the commercial use of the work is permitted. All you have to do is to take note of the license and make sure that the downloaded object is only used in the permitted range. In most cases, the creator's name has to be mentioned, too, when the object is used publicly. Depending on the license, you also have to set a link in the case of a (permitted) redistribution or representation of the work. You can view the individual license meanings at the website of creative commons (www.creativecommons.org).

Please note that the information in this section about Creative Commons licenses is not binding and cannot constitute legal advice!

3.2 Further 3D Printing Software

I will explain to you right away how the creation of a ".gcode" works. But first I want to introduce another software:

Autodesk's Meshmixer. This free software tool allows you to edit and position ".stl" files easily. Sometimes ".stl" files are not meshed correctly and contain holes or other defects.

To obtain a high-quality surface, some files must, therefore, be repaired beforehand. The process of analysis and repair can be done using Meshmixer. You can download the program free of charge at www.meshmixer.com. If you want to create your own components or realize new ideas, you have to get into the topic of engineering and will need engineering or design software. Of course, there are alternatives to the usually more expensive, professional programs such as Solidworks (Dassault Systèmes), Catia (Dassault Systèmes), Solid Edge by Siemens, or Autodesk Inventor.

As a student, you can often get such software for free or discounted as a student license. For all others, a few free programs in the following. The best approach would be to simply try a few different programs to find the right one for your purposes.

For beginners and for testing whether designing is the right thing for you, the free program Tinkercad (www.tinkercad.com) from Autodesk is recommended. Advanced users, on the other hand, should familiarize themselves with Blender from the Blender Foundation (www.blender.org) or FreeCAD (www.freecadweb.org) offered by the FreeCAD community.

Since constructing itself is a time-consuming topic and would go beyond the scope of this workshop, we will now get back to 3D printing and its software.

Let's learn step by step how to proceed with a ".stl" file, which you have downloaded, e.g., from Thingiverse, in order to get to the finished printed object so that you don't get sick of software and file formats after all this theory. As an example, we use a snap hook. Search for the term "carabiner clip" in the search function in the upper right corner of Thingiverse and select the first result displayed (see Fig. 11). Click on the button "Thing Files" and download the ".stl" file of the object with one click. You will also be shown the license conditions again, i.e., for which purposes you may use the part.

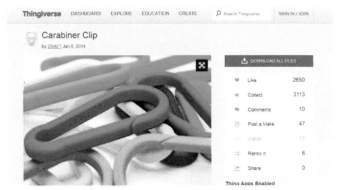

Figure 11: A snap hook designed by "ZRAFT", downloadable at www.thingiverse.com

First, we want to take a closer look at it in Meshmixer and, if necessary, repair or position the part. Start the program Meshmixer and simply drag the file into the area of the software window or click "Import". Then first click on the small arrow in the upper right corner to open the printer dropdown menu. Select your 3D printer. If it is not listed, click on "Printers preferences" and enter manufacturer and model in the appropriate fields in the next pop-up window (see Fig. 12).

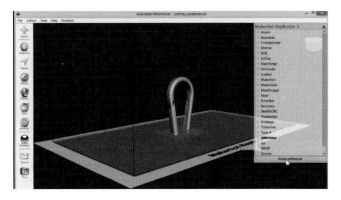

Figure 12: Meshmixer program interface with loaded ".stl" file

Also, add the printer dimensions under "Print Volume Dimensions". For the CR-10, these are 300 x 300 x 400 mm (11,811 x 11,811 x 15,748 in).

Then click on "Add" in the lower area. The manufacturer should appear in the printer list. Open the manufacturer menu in the selection tree and click on "CR-10" or select another 3D printer. Now you should have the complete installation space of the 3D printer available in Meshmixer. This setting has to be made only once at the beginning.

In order to move within the program, use the buttons of your mouse as follows: If you click and hold the right mouse button and move the mouse, you can rotate the view. If you click and hold the mouse wheel and move the mouse, you can move the view. And if you spin the mouse wheel, you can zoom into and out of the current view as usual.

As you can see in Fig. 13, the snap hook is located upright in the printing bed.

Printing the object in this orientation would not be very clever. A horizontal position provides better positioning for printing. Because, in a horizontal position, this part has a much larger contact area and no overhangs that would require a support structure. More about support structure later on. Now let's first position the print object in the desired way. To do this, click on the button "Edit" in the toolbar on the left-hand side and then select "Transform". Then enter the value 90 at "Rotate X". Alternatively, you could also rotate the object by using the colored arrows manually. Then click on "Accept". You will then see that the carabiner clip has been rotated 90° around the x-axis. But now it is still floating above the printing bed. To place it on the printing bed, click "Align" under "Edit".

The program then specifies the desired position on the printing bed. Simply click "Accept", and the carabiner clip will be positioned optimally (see Fig. 13 and 14).

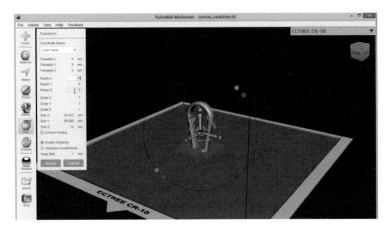

Figure 13: Rotation of the snap hook by 90° around the x-axis

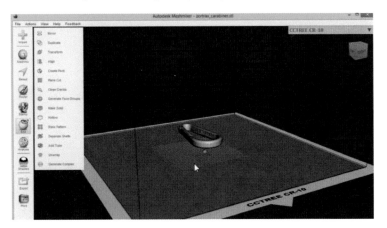

Figure 14: Result of rotation. The snap hook floats above the printing bed.

In this program, we also check the quality of the part to ensure that the surface is free of holes or gaps, which would cause defects in the printed part. To do this, click on "Analysis" in the toolbar on the left-hand side and select "Inspector". If there were holes or gaps, colored elements would appear in the area of the defects. Then you would simply click on "Auto Repair all", and the gaps in the object's mesh would be filled. Our carabiner clip, however, does not show any colored markings, so everything seems to be fine (see Fig. 15).

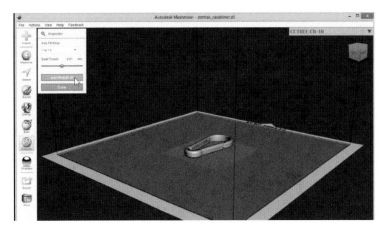

Figure 15: Checking the object with the function "Inspector"

Finish the process with "Done". A further advantage of the program Meshmixer is that you can also edit the ".stl" file in other ways. If necessary, you can cut the part in the desired orientation with the command "Plane Cut" (menu item "Edit") and remove unwanted parts of an object. Simply browse a little through the editing possibilities of the program. You will find some more useful functions for editing files. However, the presented functions should be sufficient in the beginning.

Let's save the modified file by clicking on the button "Export".

Select the file format ".stl" and the desired storage location.
Then quit Meshmixer and start Cura. The Slicing Software.

3.3 Introduction to Cura and its important Settings

First, we have to configure the slicing program to our 3D printer and make some general settings. When you start the software for the first time, a pop-up window named "Add Printer" will appear. If it does not, click on "Preferences" and "Configure Cura" in the upper menu bar. At "Printers" select the option "Add Printer". In this window, search for your printer. If you do not have an Ultimaker, switch to "Other". You can then choose your 3D printer out of a long list. With "Add Printer", you add the printer, and Cura loads the corresponding 3D printer setup (see Fig. 16).

Figure 16: Ultimaker Cura "Add Printer"

Afterward, select the desired language and currency settings in the general settings (see Fig. 17). Switch to "Printers" again and check the device settings for the 3D printer by clicking on "Machine Settings". The correct installation space, i.e., width, depth, and height, is very important. Please refer to your printer's operating manual for these dimensions.

Under the menu item "Materials" you can additionally define the filament costs and the weight of a spool. Finally, switch to "Profiles" and import a special printer profile for your 3D printer (e.g., from a 3D printing community group) or leave the default settings of Cura and select a default profile. Finish the settings with "Close".

To make all settings effective, we have to restart Cura first.

Figure 17: Defining the general settings in Cura

After the restart, the desired language setting will also be applied.

In the next step, you can check if the desired material (PLA) and the desired profile are selected in the menu by clicking on the button "Custom" in the dropdown menu under "Print setup". Then select "Expert" by clicking on the small icon next to the search bar to make a larger selection of setting options visible. In this workshop, we mainly use the print setup: "Custom" (see Fig. 18). I will briefly explain the most important settings in the following. Detailed instructions will follow step-by-step in further chapters.

If you open the "Quality" menu item, you can set the desired layer thickness of the entire print object and the first layer. Under the button "Housing", you can adjust settings for the outer shell of the print, i.e., for the "walls" and "lower" and "upper layers". Under the menu item "Filling", the "Fill density" of the print object as well as the desired "Fill pattern" are listed. In the next section, you will find all the settings regarding printing material, especially those concerning temperature. You can also set the movement speed of the walls and the filling here. We skip the points "Movements" and "Cooling" and move to the menu point "Support structure". All settings for print objects with overhangs can be made here. In the field "Printing plate adhesion", you can finally set the desired "Printing plate adhesion type". All other functions are skipped for the time being.

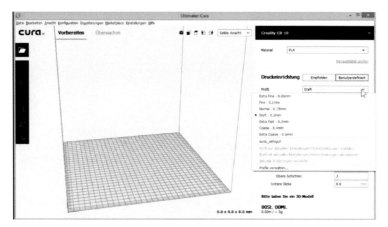

Figure 18: Loading the printer profile in the section "Custom".

Don't worry if you're a little overwhelmed by the abundance of setting options. In further chapters of this workshop, you will learn in detail how to handle these settings safely.

The movements of the view are handled in Cura in the same way as in Meshmixer. A short repetition: If you click and hold the right mouse button and move the mouse, you can rotate the view. If you click and hold the mouse wheel and move the mouse, you can move the view. And if you spin the mouse wheel, you can zoom into or out of the current view.

Now we simply drag and drop the previously edited carabiner clip (".stl" file) into the window area of Cura, and it will be placed on the virtual printing bed in the position prior defined in Meshmixer. In addition, if you have selected the "Move" button in the toolbar on the left side, you can move the part on the printing bed by clicking and holding the left mouse button. You can also use the toolbar to make some basic settings such as "Scale", "Rotate", and "Mirror", or set "Support Structure Blocker".

Correct, you can also load ".stl" files immediately into the slicing program Cura and make rotations or scaling without using Meshmixer. Since Meshmixer offers a lot of additional editing functions, I usually check and position the object in Meshmixer first. Theoretically, you could, of course, skip this step.

Now to the actual slicing. On the right side at "Print setup," we first adjust the settings individually to the desired object. Since each object has a different geometry and, therefore, different requirements for slicing, a few settings have to be adjusted almost always. However, most settings can often be kept identical for each object. If you have not already done so, switch to the "Custom" view.

First, we select the layer thickness in the menu item "Quality". Since the carabiner clip is relatively small and we want to achieve a very good surface quality, we select a layer thickness of 0.12 mm. Always choose layer thicknesses in 0.04 mm steps, e.g., as the smallest layer thickness 0.12 mm, then 0.16, 0.20, etc., when using a 0.4 mm nozzle. The higher the layer thickness, the coarser the surface, but the faster the print will be finished. Just try it, first print with 0.12 mm and then with 0.24 mm, and compare the results. Ultimately, it is always a matter of comparing the two criteria of surface quality and printing time.

Since the object to be printed is positioned horizontally on the printing bed, we also want to obtain the surface pointing upwards as high-quality as possible. For that reason, we check the "Activate Ironing" box in the menu item "Housing". A submenu will open.

All other settings can be left as they are. Also, the skipped sub-items in "Quality" and "Housing" can usually be left at the set values. The "Ironing" function is only beneficial if the surface of the print facing upwards is a reasonably flat surface. Now you may be able to see why some settings such as these always have to be adapted to the respective print object.

Then we set the desired "Filling density" in the menu item "Filling". 100 % filling means that the object will be printed completely solid. This is usually not reasonable due to time and material considerations. With very small or thin parts, it can be quite meaningful, however. Normally, filling densities of about 20 % are sufficient to obtain high-quality components. Sometimes you need a little less, sometimes a little more filling. The higher the filling density, the more compact the component becomes. In the case of the snap hook, we, therefore, choose 50 %, because we want to obtain a strong component that remains resistant under load. Since the component is relatively small, the higher filling density is justifiable. Now we could select a filling pattern. By default, you can use the triangle fill pattern, but you should try other ones as well. In a further chapter, we will deal with all of the filling patterns in more detail. All further settings up to the menu item "Support structure" are left alone. The menu items "Support structure" and "Printing plate adhesion" are two vital setting options.

With this component, however, we do not need a support structure as there are no overhangs.

With other geometries, on the other hand, support structures for overhangs are required so that the printer has an auxiliary platform on which it can build up to at a superior level. This is because the printer is unable to print in the air but always needs a previous layer on which it can build up the next layer - starting with the first layer on the printing bed. Do you remember the underlying principle of the layering process? If not, please take a look at the previous chapter of the workshop again.

How and whether support structures are needed will be clearly explained to you later on in this book. But first, we don't have to activate this function yet.

With the menu item "Plate adhesion," we set the function "Skirt" at the option "Plate adhesion type". This means that two lines are printed around the print object at a defined distance before actual printing begins. This has the advantage that first material irregularities are compensated at the beginning, and thus, the actual object can be started cleanly after a few test strips.

Since the center of gravity of this part is relatively low, and it has an adequate contact surface to the printing bed, we do not need any otherwise plate adhesion. The printing plate adhesion type "Brim," on the other hand, would be set for parts that have only a small contact surface on the printing bed and/or a high center of gravity. The "Brim" setting would print a flat first layer surrounding the print object, which increases the contact area towards the printing bed and thus prevents the part from loosening or falling. Warping would also be prevented. A printing error in which corners of the object detach from the printing bed is pulled upwards.

Now all important settings have been adjusted, so we click on the button "Prepare" in the lower right corner. The software will then perform the slicing and will show us data such as estimated printing time, required material quantity, and costs after completion. "Slicing" creates the ".gcode", i.e., the machine code for the printer. You can transfer the ".gcode" file to the 3D printer after saving it. With a ".stl" file, the 3D printer can't work with.

Experience shows that the printing time is usually a bit longer than the theoretically displayed value: 10 - 20 % of the indicated duration you have to plan additionally. In the upper area, we can finally display different views of the object by using the small dropdown menu. Instead of the standard "solid view," a kind of "x-ray view" or, very helpful for displaying support structures, the "layer view" (see Fig. 19).

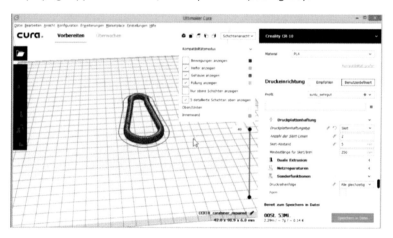

Figure 19: The sliced carabiner clip ("layer view ")

Finally, click on the "Save to File" button, select the ".gcode" format in the pop-up window, and save the file.

Well done, now you have learned all the necessary software tools to create a useful and high quality printed object out of a ".stl" file. Once you have printed the carabiner clip and removed it from the printing bed, you will hopefully be curious about what you could print next. In the following chapter, you will learn some more advanced slicing program settings and how to print more complicated geometries and larger objects. Stay tuned; you'll see it will be worth it.

4 Advanced Settings and further example Objects

4.1 Upgrade Parts for the CR-10 3D Printer

Next, we're printing some upgrade parts for our 3D printer. If you have purchased a newer model of the CR-10 series or another printer model, you can at least skip printing the parts of this chapter.

However, the content of the chapter is interesting nevertheless, because some important explanations about the placement of parts in Cura are given. So take a look at it.

It makes sense to stabilize the cable of the heating bed of the CR-10 so that it does not break off or come loose after multiple printing jobs. Therefore we will print a small device for this purpose. Furthermore, we will print four adjustment wheels for the printing bed in order to replace the smaller, originally installed ones by larger ones.
In addition, we print improved mounts for the z-axes, which come a little bit loose out of factory, and as the last part, a retainer for larger filament spools.

If you'd like to change something else or add more upgrades or modifications to your printer, just browse Thingiverse for "CR-10 upgrades" or "CR-10 modifications," and you'll quickly find what you're looking for. In my opinion, the upgrades presented here are sufficient for beginning. First, we download the ".stl" files from Thingiverse. Search for "CR-10 Ultimate Leveling Knob", "CR-10 Wire management", "CR-10 z-Axis leadscrew mount" and "creality CR-10 Spool holder" and download the respective ".stl" files (see Fig. 20). Thankfully, as with so many other files, members of the 3D printing community have already taken care of the engineering and make them available to us for free. What a service!

Figure 20: CR-10 upgrade parts you should start with

After the download, we first load the file of the adjusting wheel into Cura. Since I have already checked the files in Meshmixer beforehand, we do not need to use Meshmixer for these. As we need four adjustment wheels, we have to multiply the file first. We can do this by right-clicking on the selected model and then selecting "Multiply selected model". Alternatively, we could, of course, simply drag and drop the file into Cura three more times. If we look at the models, we notice that there are red markings on the lower side (see Fig. 21). On the one hand, this visualizes the contact to the printing bed; on the other hand, missing contact because of overhangs is shown. The small knobs of the wheels, for example, contain overhangs. From a certain degree, overhangs must be supported with a so-called support structure. This is printed with the same material, but in thinner form and with some distance to the model, in order to be removed later on by hand easily.

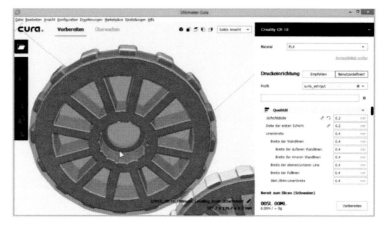

Figure 21: View at the bottom of the object in Cura

In order to avoid support structures and overhangs saving material and time as well as achieving a good surface quality, it is, therefore, best to consider an optimal positioning of the object to be printed in advance. In this case, however, we cannot significantly improve the position. And since the overhangs at the edge are only relatively small, we could simply try printing without any support structure.

In some cases, as with very small overhangs or overhangs in the range up to approx. 60 degrees, this is possible (this is also depending on the 3D printer model).

If we rotate the wheel 180° around the red axis and look from below at the area of the wheel hub, we see that this positioning would also contain overhangs.

That means that this placement would be much worse than the previous positioning since a support structure would be required in this area.

Therefore we reverse this step and keep the part in its original position. Next, we arrange our components in some desired sequence and take care of a few individual slicing settings.

First, set the layer thickness: Here we choose, e.g., 0.2 mm, then set the filling: e.g., 25 %. Fillings below 10 – 15 % are usually not useful, as they lead to fragile components and can have a negative influence on the surface quality. We choose the "Ironing" function to obtain a very smooth surface and try to print without any support structure; therefore, we do not check this option.

Then we can click on "Prepare" to slice the part and create the ".gcode" for the printer. The last step is to save the file. To create space for the next objects and to delete the previous parts, right-click on the printing bed and select "Clean printing plate".

With the following part, the z-axis mount, we can get deeper into positioning once again. After loading the file, we can see that the object is positioned upside down. If we wanted to print the part in this way, we would have to add a support structure to support the red overhangs of the two shoulders and the holes.

Now it makes much more sense to rotate the part around the green coordinate axis by 180° (see Fig. 22).

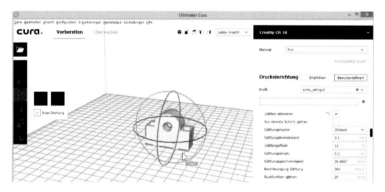

Figure 22: 180° rotation of the z-axis holder around the green axis

Because then we only have to support the upper hole and are able to avoid a lot of support structure. If we now rotate the part again by 90° around the red axis, we can ensure that the holes are printed optimally, and we can spare further support structures. However, it is usually not possible to completely print without support structures in the case of more complex parts. We then duplicate the object, because we need two pieces for two axes and then load the strain relief for the heating bed cable including the cover and the spool holder into our printing bed. These are already rotated in the right way, and we can make our settings adjustments after further desired positioning.

Since we leave the settings identical to the previous slicing job, we only need to add a support structure. So we check the "Generate support structure" in the menu item "Support structure". In general, it is recommended to place the support structures at an angle starting at 60°. The CR-10 usually handles overhangs with a smaller angle without any additional support structure on its own. Due to our adjustment, all parts on the printing bed with overhangs from 60° upwards are now supplied with support structures. We can leave all other settings as they are. We then click on "Prepare," and by selecting the "Layer View" mode, we can first take a look at the support structure in light blue. For this, the option "Show helper" must be checked. On the other hand, we can also see the printing plate adhesion mode "Skirt". And if we move the slider on the right side in order to scroll through the layers, we can examine the inner filling and the pattern of the filling prior to printing (see Fig. 23). After printing and installing the upgrade parts, we will focus on the next objects.

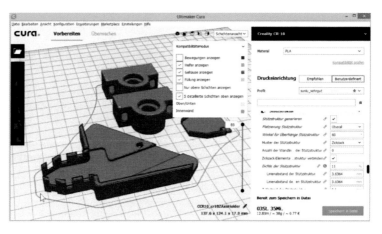

Figure 23: Arrangement of the upgrade components on the printing bed

4.2 Slicing and printing of more complex Objects

In the next section, let's get to know individual settings and a few special functions of Cura in more detail. We'll do that by looking at some more examples that we want to slice and print.

4.2.1 Low-Poly-Dog Sculpture

A low-poly-dog sculpture is just right for our purposes to deal even more closely with the topic of the support structure. Download the corresponding ".stl" file from www.thingiverse.com. You will find it by using the search term: "Low Poly Dog" (see Fig. 24).

Figure 24: Low-Poly-Dog sculpture designed by "AndrewSink" (www.thingiverse.com)

Then drag and drop the file into Cura. When we look at the bottom of the model, we see the red markers on the object that show us the areas with dangerous overhangs (as in Figure 25). To support the head and belly of the dog, we check the "Generate support structure" option.

FDM 3D printers using only one nozzle, such as the CR-10, are only able to print the support structure in the same material as the actual object. However, to make the structure easy to remove, it is printed thinner than the rest of the object. In addition, a small distance is planned between the model and the support structure. Thus, it is usually quite easy to remove the support structure.

Using a FDM dual-extruder or other printing methods, however, the support structure can also be printed from another material, e.g., a water-soluble material (e.g., PVA).

This offers the advantage that the support structure can be easily washed away after printing, and the surfaces in this area will be of higher quality.

But back to our dog sculpture. We want to generate the support structure at all endangered areas, so we choose the option "Everywhere". If you don't want to build up a support structure on parts of the model, but only where the structure touches the printing bed, you would select "Touch printing bed" instead.

If, for example, there was another element with an overhang positioned on the back of the dog, this element would also have to be supported. In this area, however, the support structure could only be built up from the back of the dog.

We place the support structures again only in areas larger than 60°. All other settings can generally be left at the set values. With the function "supporting roof" or "supporting floor," a thick material layer between model and supporting structure, either in the upper or lower area, will be modeled. And with the "Pillar" function, individual small overhangs can be supported much better. In general, I have not activated most of these functions.

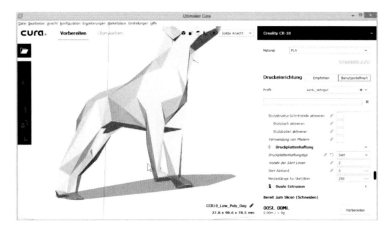

Figure 25: Bottom view of the "Low-Poly-Dog" in Cura

If you don't want support structure to be attached explicitly at certain points, but in the remaining area, so-called "support structure blockers" could be set on the 3D model. This must be done manually by using the toolbar.

Prior to slicing, we should turn our attention to the adhesion of the object to the printing bed first. The contact points (paws of the dog) towards the printing bed are relatively small in relation to the object (see Fig. 25).

So that the object does not detach during printing, we want to enlarge this contact area. This is done with the setting "Brim" at "Plate adhesion type". With this function, one layer is printed around the model in connection to the model. This increases the contact area. The width of the element depends on the object and can be freely selected. After a click on "Prepare," the selected settings of "Support structure" and "Printing plate adhesion" are displayed. And by using the "Layer view," it is possible to check the model before we start the printing process. If the option "Show helper" is checked, we can see the support structure in light blue (see Fig. 26). The "Brim" element is also displayed in light blue in the printing bed area. All dangerous areas are now sufficiently supported. Let's also try the "support structure blockers," for example, in the area of the dog's belly. After setting the blockers, slicing the object, and selecting the "layer view," area with missing support structure can be identified. However, since we do want support structure, please undo this process.

Finally, we can adjust the settings already discussed, such as "Layer thickness" and desired "Fill density," and consider whether it makes sense to activate the function "Smoothing" or not. In this case, actually not, because the dog has a slightly humpy back and head area. The nozzle could not smoothen very much in these areas, since only a flat layer, preferably parallel to the printing bed, can be successfully smoothed. So we skip this function and start the printing job after slicing and saving.

Figure 26: "Low-Poly-Dog "after slicing in layer view

When the dog sculpture has been printed, we have to remove the support structure from the printed object manually. Be careful in order not to damage the printed model. The easiest way to remove the structure is to use a selection of different pliers. Grasp the support structure as wide as possible and in one piece and remove it by bending it sideways while pulling at the same time. Any remaining parts can also be removed with a rasp or spatula. In general, the surface quality of the model is not very good in the area of support structures. Therefore, support structures are best placed in inconspicuous or not immediately visible places. This can usually be achieved by well-thought positioning of the print object on the print bed.

4.2.2 Honeycomb Vase

Next, we want to take a look at another special feature of Cura. A function that is extremely useful when you want to print a hollow object, such as a vase. In order to learn how to use this feature, we will print a vase designed by using a honeycomb pattern (see fig. 27).

For printing vases, Cura offers the "vase mode," which is hidden within the function "Spiralize the outer contours" in the menu item "Special functions". Download the file from Thingiverse (search term "curved honeycomb vase"; the file with the extension "_spiral-vase.stl").

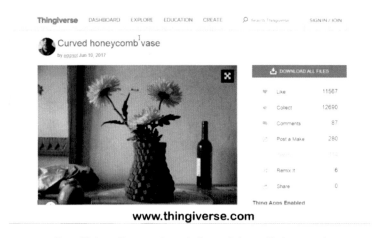

Figure 27: Curved honeycomb vase by "eggnot" (www.thingiverse.com)

When loading the file into Cura, the red markings in the area of the outer shell of the vase attract attention first. This indicates overhangs. In this case, the overhangs are very small, so we try to print without supporting structures. Furthermore, it is obvious that the file does not actually represent a vase because the object is not hollow. But that doesn't matter, because we use the special vase function of Cura. This function hollows out the object later on and transforms the outer shell of the model into a single wall. What this means is easy to see (see Fig. 28).

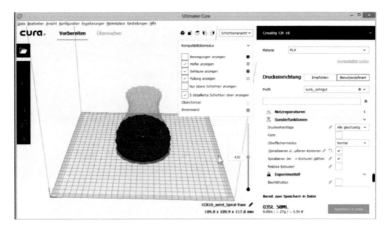

Figure 28: Sliced vase with activated function "Spiralize outer contours "

Check the "Spiralize outer contours" box and slice the model. If we then click on "Layer view," we can see - by using the slider - that the object will be printed hollow and consists only of a base and a single wall layer. That's precisely what we want it to be.

So you can save and print the object (see Fig. 29). Relatively large objects can be printed with this function in a comparably short time.

One more hint: If you don't want to place synthetic, but real flowers in the vase and thus want to fill the vase with water, you should download one of the other file variants instead, because the previously presented file will not be printed waterproof when using vase mode. Other file variants you can print in hollow form with solid walls without using vase mode.

Please read the description of the designer at Thingiverse under "Thing Details". In general, depending on the vase model, waterproof vases can actually be created in vase mode, but unfortunately, not by using this honeycomb vase in particular.

Figure 29: Almost completely printed honeycomb vase

4.2.3 Various Filling Patterns

Next, we want to test several filling patterns and get to know some more special functions of Cura. Therefore we download a quite simple cube from Thingiverse. You can find it using the search term "20 mm test cube". First, we drag the cube into Cura and copy it 12 times. Then we arrange the copies in a preferred order. For each of these cubes, we now want to apply a different filling pattern. Therefore we use the function "Settings per object" (see Fig. 30).

We select the setting "Fill pattern" for each cube and select a different fill pattern one after the other for each object. All other settings, if not selected, apply to all cubes. Then we slice the objects. With a click on the "Layer view," we display the layers to look inside the cubes. With the slider, we can navigate through the layers and view the fill patterns. It is difficult to say which fill pattern is generally the most useful for which object because it depends on several parameters (see Figure 31).

Figure 30: "Settings per object "in Cura

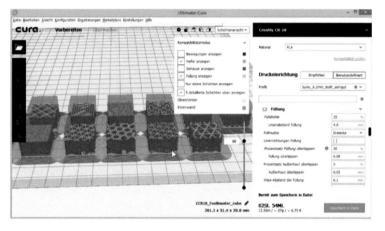

Figure 31: Layer view of the cubes, each cube with different filling patterns

In the case of simple parts, which do not fulfill any major functions or are not exposed to stress, a very simple and quick to print filling pattern can be used. The printing time also differs depending on the type of pattern. For components with defined loading directions, it is best to select a filling pattern that is aligned approximately in the loading direction in order to reinforce the structure of the component.

In case of doubt, however, a simple triangular pattern can be used as a standard. If the visualization in Cura is not enough, we can, of course, print the cubes with the different filling patterns and analyze them for further properties. To print only the filling patterns, we have to remove the outer walls and the upper layers first. We do this by replacing the set value for "Number of wall lines" and "Upper layers" with zeroes at the settings for all components. The fields in Cura will then be marked as this does not normally represent reasonable settings.

After slicing, we can take another virtual look at the result before printing. It is also best to note at which position which filling pattern was set, so that you can still keep track of it after printing. Instead, you can also save the Cura project.

You can use this procedure for other printing parameters as well. For example, if you want to change the printing profile in order to find an optimal value for a profile setting.

4.2.4 Further Tips for Slicing

Now you can consider yourself very familiar with the functions and settings of Cura.

Some final hints:
Experiment with the following parameters if you want to try to reduce printing time or to change surface quality. For printing time, apart from quality, i.e., the layer thickness and the printing speed is also relevant. But be careful, if you print to fast, you will usually have to accept a loss of quality.

In addition to the layer thickness, as the main parameter responsible for the surface quality, wall thickness, and the number of layers, as well as the "Ironing" function, are also responsible for surface quality. You can experiment with these settings if you want to make changes here.

For saving a modified profile for later use or cataloging, click on "Profile" in the dropdown menu and "Create profile of current settings/overrides".

Meanwhile, you have acquired a solid and advanced knowledge of necessary 3D printing software and can confidently try other objects and especially larger or multi-part print models.

We will now turn to various filaments and materials in the concluding chapters. There is also an additional chapter on printer maintenance and troubleshooting for fixing poor printing results.

Finally, 3D scanning is presented. So be curious.

5 Filaments and Materials

So far, we have only printed with PLA or PLA+. These materials are probably the easiest and safest to use for FDM 3D printing. However, there are many other materials that can be used for printing—for example, ABS, PETG, and TPE, or TPU. The special thing about TPU and TPE is that they are flexible plastics. Of course, this opens up completely new ways. You can print yourself a bathing duck or a flexible cover for your smartphone. Even if flexible plastics, like most of the other materials than PLA, are more difficult to print, with a little experimentation and patience, it should be quite possible.

Figure 32: A selection of various 3D printing filaments. Different materials and colors

ABS, one of the most widely used plastics, is one of the stiffer 3D printing materials. The advantages of this material are the high achievable values for stiffness, toughness, and strength, i.e., the mechanical properties.

The negative aspect of ABS, on the other hand, is the odor generated during printing, which contains toxic substances and should not be inhaled. When printing with ABS, it is, therefore, essential to ensure adequate ventilation.

As a general rule, ventilation should also be provided as a precaution when printing with other materials.

PETG material is a good alternative to ABS and PLA. On the one hand, it is as easy to print as PLA; on the other hand, it is usually odorless and very tough. PETG filament should be used, especially for components that need to have weather-resistant properties. PETG is, by the way, a modified form of PET, the material most one-time-use deposit bottles are made of.

The easiest way is to test all materials on a small scale to familiarize yourself with them. For this purpose, you can often order small test spools, which are already available for $ 5 - $ 10.

In addition to standard materials, there are also more exotic material mixtures. For example, wood filament: A filament, usually made of a PLA based material, to which real wood fibers have been added. It gives the filament a natural and wood-like character. This can also be detected by the smell during the printing process. This material is ideal for life-like objects. For example, you could print small trees for a model railway or an architectural model. The difficulty of 3D printing with wood filament is that the print nozzle can be clogged relatively easily by the wood particles.

There are also filaments that are mixed with metal powders, e.g., silver, aluminum, or copper. And there are also filaments with carbon fibers. Since other particles, such as conductive particles, can also be added to filaments, objects with integrated conductor paths can be printed. The "conductive" filament of the manufacturer "Proto-pasta" has to be mentioned. You may be able to imagine the dimension and potential of this. Who knows, maybe in the distant future we will simply print our new smartphone by ourselves at home.

The filaments with metal powder additive you could also use for jewelry. Try printing a ring or necklace!

That's all regarding filaments, experiment a little bit, this is the best way to familiarize yourself with the possibilities and materials. It is also advisable to get a smaller or larger selection of different colors. Especially for print objects that are composed of several parts, it can make the object look like a lot more creative if you print these parts in different colors.

One final note: Remember that support structure waste or poor printing objects have to be disposed properly. It is best to contact the manufacturer or your local disposal association for further information.

Nowadays, there are also recycling machines, such as the Protocycler from ReDeTec (www.redetech.com). This machine can shred and meltdown 3D printing filament leftovers and rewind "new" filament onto an empty spool using an integrated recycling process.

However, these systems are often not yet profitable enough for users with small amounts of leftover filament. The cost of such a machine is about $ 1500.

6 3D Printer Maintenance and Troubleshooting

6.1 Repair, Maintenance, and Troubleshooting

To enjoy your printer for a long time, it is a good idea to check all bolts for tightness every now and then (see Fig. 33).

Figure 33: Checking bolts for sufficient tightness

After 10 - 20 printing operations or if there are discrepancies with adhesion, the printing bed calibration should be redone. If a component of the 3D printer has been damaged, there are usually spare parts or repair services. You can find helpful video instructions on the Youtube channels of some manufacturers for the exchange of parts. From time to time, the nozzle should also be cleaned from filament residues, especially if it is clogged. In the following, a short description of the cleaning process: Heat up the nozzle of the printer manually. Remove the filament and clean the nozzle using an acupuncture needle or a similar thin piece of wire.

The printing bed can be cleaned best with a microfiber tissue using isopropyl alcohol or conventional part cleaner. A part cleaner can be found, for example, in the nearest DIY store. If you accidentally damage a print, for instance, when removing it from the printing bed, this is especially annoying for longer lasting prints. Since this has happened to me from time to time, I prefer to use some glue in such cases. Especially with smoothly broken off parts, this small cheating can usually hardly be recognized afterward and is a quite stable option. A good glue is basically any kind of superglue (see fig. 34). Voilà, the print is saved!

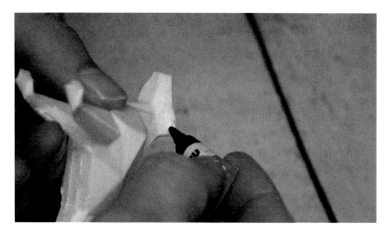

Figure 34: Refix damaged 3D prints with superglue

Another problem is the planning of the remaining amount of a half-used filament spool. Is it enough to fully print my part, or do I run out of material during printing? This is a tricky guess sometimes. If you prefer to be safe, you can weigh the filament on the spool. Remember that you have to subtract the weight of the empty spool. The weight of the print object is displayed next to the printing time in Cura, so you can compare whether there is still enough filament on the spool or not. It is better to plan a little more generously. If the filament should run out during printing, you can also change the filament during printing. However, you will probably see a small flaw in the finished print at this position. To change the filament during printing, do the following: Navigate to "Pause Print" in the printer menu. Then pull the remaining filament out of the nozzle and change the filament spool. After inserting the new filament, start printing again by selecting the "Resume print" function. The print head starts moving again and completes the rest of the printing job.

Some 3D printers are equipped with a filament sensor. This automatically stops printing if the filament runs out during printing.

6.2 Troubleshooting Guide

Let's get to the actual troubleshooting guide. This chapter shows you some common printing errors and how to fix them. It is intended to help with poor 3D printing results.

6.2.1 Poor Adhesion

The whole print or the first layer does not adhere to the printing bed. This problem is relatively essential as the entire printing process will not work if the first layer is missing. So you can stop the printing process after the first few minutes right away. The most common cause of this problem is that the printing bed is not properly leveled, meaning that the nozzle is too far away from the print bed. Carry out the printing bed leveling one more time and reduce the distance from nozzle to printing bed via the adjustment wheels of the bed. Other causes may include excessive printing speed or incorrect temperature settings of the print bed or filament. In most cases, the temperature is too low. If the contact surface is very small, the "Brim" function in Cura may help. This can be found in the menu item "Plate adhesion" at "Plate adhesion type".

6.2.2 Under-Extrusion and Over-Extrusion

An FDM 3D printer usually has no feedback function to tell the software whether there is enough material coming out of the nozzle or not. If not enough material is extruded, but the printer is not informed about that, the error pattern of under-extrusion will occur (holes, gaps, or thin structures), which means that too little filament is processed. A typical cause is often a wrong value for the filament diameter in the slicing software. Check whether the correct filament diameter is entered, e.g., 1.75 mm, when using the CR-10.

Over-extrusion, on the other hand, is exactly the opposite of under-extrusion. This means that too much filament flows out of the printer's nozzle. This problem can usually also be traced back to an incorrect value in the "Filament diameter" setting in the slicing software.

6.2.3 Stringing or Oozing

Stringing, also called oozing, causes the formation of spider-like structures around the print object. This is usually caused by residual material being extruded out of the printer's nozzle while the print head is already moving to the next point. In Cura, there are two important settings for solving this problem. The first is the setting "Activate feed" located at the "Material" menu item. Here you can increase the feed distance and feeding speed. On the other hand, the setting "Activate Coasting" in the menu item "Experimental" is useful. Check both functions alternately and compare the results.

6.2.4 Layer Separation

Layer separation, also known as Layer splitting, occurs when the individual print layers cannot connect properly. Then, crack-like conditions occur in the object. One of the reasons for this is, that the layer thickness is too high. In general, the layer thickness of the print object should be 20 % smaller than the diameter of the print nozzle.

That means with a 0.4 mm printing nozzle, such as the one installed on the CR-10 series, the layer thickness should therefore not exceed 0.32 mm. Generally speaking, smaller layer thicknesses achieve the best results in terms of quality. Another possible cause is that the temperature setting is too low. A suitable melting temperature should be selected for each kind of material individually so that the plastic can melt properly and also bind well with the previously printed layer. These options can be set in the settings for the printing nozzle. You can also download a so-called "Temperature Tower" from www.thingiverse.com to find the optimum temperature settings for the respective filament. The ".gcode" of a Temperature Tower will lower or raise the temperature step by step during the printing process, and the individual sections can thus be easily compared with each other.

6.2.5 Clogged Nozzle

If no filament or only a small amount of filament is being extruded, but all other functions are working as desired, the nozzle is most likely clogged. There are several ways to solve the problem. On the one hand, remove the filament, heat the nozzle manually, and clean it. On the other hand, you can also replace the nozzle in case of heavy wear or clogging. A new nozzle is available for just a few bucks or is supplied as a spare part delivered with the printer. If the whole "hot-end" is affected, it becomes more difficult. The hot-end is the area where the filament is melted. In this case, only intensive mechanical cleaning helps.

6.2.6 Poor Infill

If the inner structure of the object, i.e., the filling, is printed in very poor quality, this can have several reasons. On the one hand, you can try a different filling pattern; on the other hand, increasing the filling density to more than 20 % in the "Slicing" settings is usually helpful. If this alone does not lead to the desired result, the problem can also be due to a too fast printing speed of the filling structure. Try a lower value instead.

6.2.7 Vibrations and Ringing

If wave-like structures appear on the outer shell of the printed object, this is probably due to a printing speed that is too high. Even if it increases the printing time considerably in some cases, try a lower value, it will convince you by much better surface quality. Other causes are mechanical problems. Tighten all bolts and make sure that the belts of the stepper motors are properly tightened.

6.2.8 Warping

Sometimes a corner of the print object may come off the printing bed and deform upwards during printing. This phenomenon is known as warping. The solution is an adequate printing bed temperature with PLA, e.g., between 50° and 60° Celsius. If you cannot solve the problem in this way, also try to select the function "Brim" under "Plate adhesion type" in the "Slicing" settings. In most cases, this will lead to the desired result.

6.2.9 Scars on the Print

If the nozzle moves in areas that have already been printed, this may reduce the number of required filament feeds. This function is called "Combing". On the other hand, the hot print nozzle, can melt the plastic of an already printed layer in this area, or extrude some material within the bottom layer of the object and thus leave traces. In Cura, you can activate the function "z-jump on retraction" in the menu item "Movements," deactivate the "Combing" mode, or set it to the setting value "Inside filling".

If your printing problem is not listed here or if you have another, more specific problem, you can also search for solutions at other Troubleshooting guides on the internet. There is also the possibility to post a photo of your specific printing problem in a 3D printing help forum or in 3D printing groups of social networks to get individual help and advice.

7 3D Scanning

Coming to the end of our workshop, I would like to present the method of 3D scanning!

Besides the more expensive 3D scanners on the market, like Makerbot's Digitizer, which uses two lasers to scan the object on an automatic turntable, or the Shining 3D "EinScan" platform scanner, there are also 3D scanners in the form of handheld devices, like the 3D scanner from xyz-printing or the "Sense 2" from 3D-Systems. These hand and platform 3D scanners range from 200 to several thousand euros.

A tip: If you own a Microsoft Xbox 360 with so-called Kinect sensors, you can also use suitable software, e.g., "Skanect Free" or "Skanect Pro", to turn it into a fully functional 3D scanner.

But my favorite 3D scanning method is definitely one of the easiest and cheapest. Those who accept some compromises in scan quality need nothing more than a smartphone, an app, and a piece of paper for their first 3D scan. This is possible by using the smartphone app "Qlone".

You can download this app for free from an app store that is suitable for your smartphone. The scanning process itself is also free of charge. Costs arise only if a scan process is then saved, for example, as a ".stl" file. For each saved scan, then approx. $ 1 in costs result. A bargain in comparison to the more expensive devices, in particular, if you want to scan only occasionally.

To scan an object, you also need a special pattern, which you can download free of charge from the manufacturer's website (www.qlone.pro). The scanning process proceeds as follows: You place the object to be scanned on the grid pattern and then move your smartphone at different heights in a circular movement around the scanning object in order to completely fill the virtual half-sphere displayed. Alternatively, you can place the scanned object on a turntable and rotate the object around the camera. In this way, you may achieve an even better scan resolution (see Fig. 35).

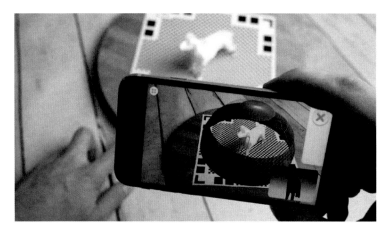

Figure 35: 3D Scanning using the app "Qlone".

When the scan process is finished, you can decide whether you want to save the scan as a ".stl" file or want to discard it (see Fig. 36). The ".stl" file can then be edited in Meshmixer and sliced into a ".gcode" using Cura.

Figure 36: After the scanning process a preview of the file is provided

This concludes our 3D printing workshop, but now it's even more important to get in touch with your 3D printer and materialize many great ideas. Browse through community platforms like Thingiverse or be creative yourself, there are a lot of options!

If you did enjoy the workshop, please recommend the book to your friends and become a part of the growing 3D printing community, and please write a short positive review of this book. That would mean a lot to me and helps me to get some feedback! Thank you very kindly!

Figure 37: Good luck and enjoy 3D printing!

Made in United States
Orlando, FL
30 November 2021